David Williams Cheever

Two Cases of Oesophagotomy for the Removal of Foreign Bodies

With a History of the Operation. Second Edition

David Williams Cheever

Two Cases of Oesophagotomy for the Removal of Foreign Bodies
With a History of the Operation. Second Edition

ISBN/EAN: 9783337326487

Printed in Europe, USA, Canada, Australia, Japan

Cover: Foto ©berggeist007 / pixelio.de

More available books at **www.hansebooks.com**

TWO CASES

OF

ŒSOPHAGOTOMY

FOR THE

REMOVAL OF FOREIGN BODIES:

WITH A

HISTORY OF THE OPERATION.

SECOND EDITION, REVISED, WITH AN ADDITIONAL CASE.

By DAVID W. CHEEVER, M.D.

ADJUNCT PROFESSOR OF CLINICAL SURGERY IN HARVARD UNIVERSITY; SURGEON TO THE
BOSTON CITY HOSPITAL.

———

BOSTON:
JAMES CAMPBELL, 18 TREMONT ST.
1868.

DAVID CLAPP & SON, Printers,
334 Washington Street.

INTRODUCTORY REMARKS.

THE removal of foreign bodies from the œsophagus, by external incision, has been so little treated of in surgery, that it is dismissed, by most authors, with a bare allusion, or an obscure description, as of some remote possibility, which the surgeon should know of, but which he need have no expectation of being called on to undertake.

And it has been so rarely performed, that, after diligent search, we have been able to find a record of only *fifteen* authentic cases : of which *seven* were in France, *one* in Italy, *one* in Belgium, *one* in India, and *five* in Great Britain.*

The ingenuity of the surgeon has been exhausted in devising means to withdraw foreign bodies through the natural passages; an intention which cannot be too highly commended, but which has often led to irritation, ulceration and perforation of the digestive tube, by the too zealous employment of instruments, as dangerous as the foreign bodies themselves.

By reviewing the history of the operation, and by our own cases, we hope to show that, as in strangulated hernia, urinary extravasation, or croup, so in œsophagotomy, the danger is in delay. That the risks of the operation depend on skill in its performance, and not on its sequences. That if done early, it can be done safely. And that it deserves its place among operations of necessity, and not of expediency, as one which can relieve a fellow being from exquisite suffering, and from extreme peril.

---

* Since the above was first published, six new cases have been added. Of these, four are of foreign bodies ; and two, of stricture.

PREFACE TO THE SECOND EDITION.

THIS little monograph was privately printed by its author, in February, 1867. The edition is exhausted; and so many inquiries are made for it from distant parts of the United States, as well as some from Europe, that he is led to re-publish it.

The table has been enriched with several new cases, sent the author by Dr: L. Voss, of New York; and one, of original operation, by Dr. ALFRED HITCHCOCK, of Fitchburg, Mass. An additional case of the author's is given; which, while it illustrates the fallacy of diagnosis, confirms the innocuousness of the operation.

D. W. C.

1267 Washington Street, Boston,
      September, 1868.

ŒSOPHAGOTOMY.

CASE I.

Mr. L——, a young man of temperate habits, and strong, muscular frame, an engineer by profession, while eating a dinner of chowder, on the 9th of November, 1866, suddenly became conscious that he had a bone in his throat. Violent strangling and efforts at ejection came on, followed by vomiting. The fragment was not dislodged. A physician, having arrived, caused him to masticate and swallow some ship-bread, without moving the foreign body. His throat being violently congested, he was directed to go home, and dress it with cold water, and if trouble continued, to send for me. On the same afternoon I was sent for, but was not at home, and word was not left for me to visit him. Distress in the throat continued, and while repeatedly applying cold water, continued chills came on. He passed a bad night, swallowing only water. The next day, as bad as ever.

He now, Nov. 10th, sent for me again, and I saw him between 2 and 3 o'clock, P.M., a little more than twenty-four hours after the accident. He was in

much distress at each effort to swallow ; his face was
suffused ; his eyes congested ; the palate swollen, the
uvula touching the tongue, and provoking attempts
to swallow an oversecretion of mucus which collected
in the fauces.　There was no swelling of the tonsils,
or tongue.　The latter was coated ; the pulse a little
accelerated.　Pains radiated up and down the neck
from a point on a level with the cricoid cartilage, on
the right side.　He had slept none the•past night.
Having placed him opposite a good light, I explored
the pharynx with my forefinger.　I was able to reach
over and behind the larynx, and searched the base of
the fauces and the folds around the epiglottis, as well
as the back of the pharynx.　I felt nothing.　He
vomited a little mucus, and expressed himself relieved.
He now swallowed a whole goblet of water, without
much wincing.　I then left him, with directions to
rub the neck gently with a liniment containing one-
third laudanum, and to wait and see whether the distress returned.　We both of us had hopes that the
foreign body had passed down.　This was on Saturday afternoon.　I heard no more from him until
Monday morning, when I was sent for early, and
found him worse.

It seems that the distress, on swallowing, returned
in two hours after I explored the fauces on Saturday,
and had continued to increase.　The two nights
(Saturday and Sunday) were poor ones.　He could
sleep only a half hour at a time, by lying with his
mouth open and allowing the saliva to dribble away.
As soon as it accumulated, the effort to swallow woke

him up. These naps were promoted by rubbing in the laudanum, of which he had used large quantities. He had swallowed nothing but water and a very little strained broth. He could not drink milk, and could distinguish the difficulty of swallowing between it and water. His efforts to swallow were attended with violent contortions. His suffering was evident and great. His aspect was worse everyway. His tongue and throat the same as on Saturday. His pulse 116.

Having prepared a soft sponge, large enough to fill the œsophagus, and attached it to a long probang, I passed it down twice. The first time it was passed, he vomited ; the second time, he fainted and declined another trial, which, indeed, I was not disposed to make. The sponge passed down almost to the sternum, and beyond the seat of pain, without meeting any obstruction. On withdrawing it the second time a single drop of fresh blood appeared on its left side, corresponding to the *right* side of the throat. No blood tinged the mucus expectorated, before or afterwards. He continued to refer the centre of pain to the right side of the cricoid cartilage. Tenderness on pressure was greater there than elsewhere, much greater than on the other side of the neck. Pain radiated down on the chest, upwards to the temple, and from the throat to the ear. It was almost impossible for him to swallow, and he would not attempt it unless urged to do so. I now gave him some morphia, directed his bowels to be cleared by a

2

purgative enema, preparatory to using nutritive injections, and left him for a few hours.

My diagnosis at this time was, that a small fragment of fish bone was imbedded in the wall of the œsophagus on the right side, opposite the cricoid cartilage, this being the line of junction of the pharynx and œsophagus. The reasons on which this opinion was based were :—

1st. The history of the case ; it being now the third day since the accident, and the symptoms increasing in severity.

2d. The morbid appearances in the fauces, of the palate, uvula· and tongue, were not sufficient to account for the symptoms.

3d. That the foreign body was small, because the fingers, lifting forward the larynx and trachea and embracing the œsophagus, could detect nothing externally, and the probang encountered no obstruction.

4th. That it was situated opposite the cricoid cartilage on the right side, and was imbedded there :

    Because this was the narrowest part of the œsophagus·;

    Because this was the centre of pain and tenderness ;

    Because the drop of blood on the sponge corresponded to this side.

At 1.30, P.M., Dr. Buckingham saw. him with me. Since I last saw him he had had a violent rigor, followed by heat and sweating. His pulse was 120, sharp and irritable; œdema of the right side of the neck had come on. His face was haggard ; his eyes and skin much congested. He expressed the opinion that he could not endure his sufferings many nights and days longer. Dr. Buckingham wisely refrained from exploring the parts, in view of their œdematous condition and the constitutional disturbance. He concurred with me in the opinion that œsophagotomy should be attempted, as the only safe course; and that the operation was imperatively demanded and could do no harm.

1st. Because, if the diagnosis were correct, and the foreign body there, it was working through the walls of the œsophagus, and could be reasonably reached in no other way.

2d, Because, if the diagnosis were incorrect, and the foreign body gone, the irritation it had excited was about to lead to abscess, which might end in suffocation, and for which the incisions of œsophagotomy must be the ultimate treatment.

The risks and chances having been detailed to the patient, he assented promptly to the operation. He was conveyed to the City Hospital, and having been kept under the influence of morphia, and stimulated by an alcoholic enema, the operation was begun, at 7, P.M., with the very valuable assistance of Drs. Buckingham and Thaxter.

OPERATION.

The patient had a short, thick-set neck, with powerful muscles, and the veins were congested and full. The neck having been extended and the face turned to the opposite side, an incision was begun opposite the top of the thyroid cartilage, midway between it and the sterno-mastoid, and carried down parallel to the muscle, three and a half inches, to the sternum. This divided the skin, superficial fascia and platysma. The edge of the sterno-mastoid was now sought for and dissected out. In completing this dissection just above the top of the sternum, the anterior jugular vein, a vessel of moderate size, was divided. This, which would have been of no importance otherwise, gave rise to two consecutive sucking sounds, as if two bubbles of air had been drawn into the vein. Dr. Buckingham promptly applied his finger to the mouth of the vein, and no perceptible effect was produced on the heart's action by this accident. This vessel, although it did not bleed, was subsequently tied, as a matter of precaution; and this was the only ligature required during the operation.

The carotid sheath was next sought for and exposed with the point of a director, and then drawn outwards with the sterno-mastoid. Then the upper belly of the omo-hyoid muscle was denuded, and drawn outwards also. The edge of the sterno-hyoid next presented itself, the thyroid gland, and the side of the cricoid cartilage. A slow dissection with the director and the handle of the knife, and with the finger, was now carried on through the cellular tissue, down between the œsophagus and the carotid sheath, which in the lower part of the incision lay in contact, side by side. The side of the. œsophagus having been reached, the larynx and trachea were lifted and tilted over, and the foreign body felt for, in vain.

The next step was to stretch the œsophagus from the inside, so that its walls might be presented in a single layer to the finger. To do this the tube of a stomach pump was introduced into the œsophagus by the fauces. Considerable difficulty was experienced in passing the tube into the œsophagus, as it repeatedly brought up in the larynx and trachea. The swollen state of the soft palate threw the tube forward into the glottis, and the patient being profoundly ætherized, the epiglottis did not exclude it, and he breathed through it in a rhythmical manner, and without lividity. After sundry vain attempts by myself and others, Dr. Thaxter succeeded in drawing forward the opening of the glottis, and in guiding the tube, over his finger, into the pharynx and œsophagus. No obstruction was met with in the latter, and the tube was pushed down to the sternum.

A careful search with the finger now revealed a foreign body lying between it and the stomach tube. Opposite the cricoid cartilage, at the junction of the pharynx and œsophagus, on the right side, some sharp, small substance was pricking through the walls of the œsophagus, a little towards the vertebral or back side of the gullet. The depth of the wound, and the oozing of blood, made it impossible to see the foreign body, which was sought for, and ultimately withdrawn, by touch alone. The intervening portion of the wall of the œsophagus was pricked through with a director and dressing forceps, and, after several flat scales had been removed, the main portion of the bone was drawn out. It consisted of a rough, sharp-

edged, but flattened, piece of the fin-bone of a fish,
and was one-half an inch long, by one-fourth broad.
No other pieces being detected the tube was withdrawn. The wound was left entirely open, as it was
thought that it would be extremely difficult to remove
any stitches placed in the œsophagus, even if they
did good, while the freest exit should be allowed the
discharges. The wound of the gullet, being a lacer-
ated one, did not bleed. The patient·was given an
injection, subcutaneously, of one-fourth of a grain of
morphia, and removed to bed.

In reviewing the steps of an operation to reach the
œsophagus by external incision, it appears, first of
all, that this incision cannot well exceed three, to
three and a half inches, in length. This is about the
distance, in average necks, from the top of the thyroid
cartilage to the sternum. If we cut above the thyroid
cartilage we endanger the hypoglossal nerve and
lingual artery, in a deep dissection ; and, more important, the *superior laryngeal nerve.* The latter crosses
the space between the hyoid bone and top of the
thyroid cartilage to enter the larynx, and its section,
would destroy the sensibility of one vocal cord, and
one-half of the glottis. Through a comparatively
short incision, therefore, we are obliged to make a
very deep dissection, down to the prevertebral muscles,
and to draw various important structures out of
harm's way as we proceed. First, the carotid sheath,
containing the artery, vein and pneumogastric nerve,
which approximate closer and closer towards the
œsophagus, as we descend the neck. Above and

below are the superior and inferior thyroid arteries. On the inside, the thyroid gland. Below the finger, the sympathetic nerve. And, finally, running up between the œsophagus and trachea, to the back of the larynx, the inferior, or *recurrent laryngeal nerve*, the motor nerve of the larynx, whose section would paralyze one-half of the glottis; and the partial division of some filaments of which, in one case of œsophagotomy, led to a permanent alteration of the voice. This nerve lying upon the front of the œso-phagus principally, is to be avoided by opening the gullet towards its *posterior* part. The œsophagus is easier found also on the *left* side of the neck, as it naturally inclines to that side. And the rule has been laid down that œsophagotomy should be done on the left side as the place of election, unless we are sure of cutting down on the foreign body on the *right* side. The deeper dissection being carried on chiefly with the director, it is possible to reach the œsophagus not only *without injuring any nerves*, except the unimportant superficial branches of the anterior cervical plexus, but also, as we shall show in the second case, *without tying a vessel*.

For the following very minute record of the subsequent history of the case, I am indebted to my House-surgeon, Dr. J. B. Brewster, to whose devoted care the operation owes much of its success.

The patient was removed to the steam-room. At 11, P.M., he is sleeping. Respiration 30, and not very noisy. Enema of beef-tea given.

Nov. 13 ; 4, A.M.   Awake ; voice is husky ; says that swallowing is more painful than before the operation.   Wound dressed with a wet compress.

8, A.M.   Pulse 120, full ; tongue slightly coated and easily protruded.   Slight suppuration in wound ; great thirst.   To have his mouth moistened with a wet cloth, and his arms and wrists bathed freely with cold water, to relieve his thirst.   Absolute diet ; not to talk ; not to swallow, and to allow the saliva to run from mouth.   Room to be kept full of steam, both to relieve the larynx and to check the watery transudation from skin ; beef-tea enemata, ter die ; morphia, p. r. n.

6, P.M.   Can swallow a little easier ; pulse 120 ; feels quite comfortable, slept most of afternoon ; at 8, P.M., to have beef-tea enema ; at 9, morphiæ, gr. $\frac{3}{16}$th, subcutaneously.

Nov. 14 ; 6, A.M.   Slept well until 2, A.M.   Feels more comfortable, and thinks he could swallow water without pain, but is not allowed to do so.   Feels hungry, and has beef-tea ℥ ij. *per rectum*, which immediately satisfies his desire for food.   Morph. sulph. gr. $\frac{3}{16}$th, subcutaneously.

9, A.M.   Neck less swollen ; suppuration in the wound.

2, P.M.   Purulent expectoration.

6, P.M.   Patient is much worse ; no morphia since morning ; expectoration very troublesome ; pulse 120, and more compressible.   Ordered morphia and enemata.

Nov. 15 ; 2, A.M.   House-surgeon was called in

haste by the nurse, because patient breathed badly.
The difficulty was found to arise from a collection of
mucus and purulent matter in the throat.   It was
dislodged by swallowing a little water.

9, A.M.   Pulse 108; swallowed a little milk and
water without much pain ; a part of it ran out through
the wound.   *Third day since the operation.*

Nov. 16 ; 9, A.M.   Has had enemata, and morphia ;
a good night ; not much expectoration.   But little
discharge from wound, which is clean and healthy.
Ligature upon anterior jugular came away.   Pulse
112, of good strength.   Drank a little cream and
water ; most of it ran out of the wound.   To drink
nothing but water ; continue enemata.

6, P.M.   Aphthous ulceration on lips and tongue ;
ordered borax and glycerine.   Thirst continues ;
drinks much water, a part of which escapes through
the wound.   Talks more plainly.

Nov. 17; 9, A.M.   Slept well ; pulse 88, and good ;.
countenance improved ; voice stronger ; aphthæ troublesome ; wound clear and healthy.

6, P.M.   Tongue very sore ; more suppuration in
the wound.   No morphia through day, but obliged
to give it in the night.

Nov. 18.   Pulse 96 ; swallowing less painful ; not
so much thirst ; more suppuration ; several sloughs
have been discharged from the wound.

Nov. 19.   Aphthæ better ; a slight diphtheritic
exudation upon the gums.   Slept all night without
morphia.   More large sloughs from wound.   Two-
thirds of the water taken passes into stomach.

3

Nov. 20. A good night, without morphia. Profuse reddish discharge from wound. Pulse 96, compressible. Enemata continued.

One week since operation. Drank cream ℥ ij., and beef-tea ℥ iij., and but little escaped from the wound. The effort of swallowing presses out quite a quantity of dark pus, and several sloughs ; after swallowing food, felt a little sore.

The steam was shut off. To inhale twice daily chlorate potassæ, atomized. Increase frequency of enemata. To drink beef-tea, also. This P.M. had a healthy evacuation of the bowels.

Nov. 21. To have milk ℥ viii., three times a day ; but little flows through the wound. The dryness and discomfort of throat relieved by the atomizer.

Nov. 22. Pulse 72 ; mouth not so sore. Discharge from wound is of excellent character. During the night drank three tumblers of milk.

Nov. 23. Hurts less and less to swallow.

Nov. 24. Wound full of granulations, and healing at ends. No morphia for a week past.

Nov. 25. Only a drop or two of milk flows through the wound ; suppuration less.

Nov. 26. Sat up half an hour ; *two weeks since operation.* No milk came out of the wound this morning.

Nov. 27. Swallowing without pain.

Nov. 28. Takes three quarts of milk daily, and four enemata of beef-tea.

Nov. 29. Sits up half an hour, twice a day.

Nov. 30. Granulations in wound exuberant ; caustic applied.

Dec. 1. Is troubled by a tickling in throat, which causes him to swallow constantly. Upon examination, the uvula was found elongated, and the palate very pale in color. Touch uvula with a solution of arg. nitrat. grs. xxx. ad aquæ ℥ j. ℞. Ferri et potassæ tart. grs. x. ter die, in milk.

Dec. 2. When he drinks, hardly loses a drop.

Dec. 4. *Three weeks since operation. From this date the œsophageal fistula seemed definitely closed.* Patient gaining flesh and strength.

Dec. 6. Pain and stiffness in trapezius muscle. To omit beef-tea enemata; to have milk and chicken broth, *ad libitum.*

Dec. 9. Sat up four hours; went to bed on account of pain in shoulder, which seems to be rheumatic. Morphia subcutaneously.

Dec. 12. External opening of wound small, but extending into a cavity two inches deep, whence pus can be pressed out; to be syringed out daily.

Dec. 15. Pain in shoulders more severe. Ordered mustard sinapism, and morphia.

Dec. 19. Except for rheumatism, feels perfectly well. Fistula diminishing, and skin contracting; sits up all day; takes rice in his broth; oat-meal gruel, made thick, and soft toast; to have shoulders covered with flannel and sulphur.

Dec. 21. Rheumatism continues; urine examined, and found *alkaline.* Ordered, ℞. Vini colchici gtts. x.; fl. magnesiæ ℥ ss., bis die. *Eats beef-steak.*

Dec. 31. Well, so far as his œsophagus is concerned. Still troubled with the rheumatism, for which he remained in the hospital several weeks longer.

After a closure of several months the œsophagus wound re-opened. It remained open several months, and a portion of carious bone (apparently from the body of a cervical vertebra) was discharged. Slight angular curvature of the spine (about the 7th cervical) also existed. The fistula closed; and the patient is now (eighteen months since operation) well.

CASE II.

On Friday, the 16th of November, 1866, *four days only after the preceding operation*, there presented himself to me at the Surgical Room of the Boston Dispensary, Daniel B——, a young man of average strength and health. He stated that while hastily eating some cabbage at supper time, on Wednesday, the 14th, he felt some foreign substance lodge in the fauces, which, on reflection, he concluded to be a fragment of bone, boiled off from the meat with which the cabbage had been cooked. He had made many efforts to dislodge it, in vain. His nights had been sleepless; pain and difficulty on swallowing were nearly as great as in the first case. His pulse was 120; tongue foul. The sensation of a foreign body, the tenderness and so forth, were referred to about the level of the cricoid cartilage on the left side. The power of locating the exact spot where the foreign body might be, was not so clear as in the former case. The patient gave signs of severe suffering on trying to swallow. He had taken no food since the accident occurred. An inspection of the fauces revealed nothing of importance.

A sponge probang was gently passed down to the sternum. No obstruction was felt. No blood was brought up on the sponge. But the first expectoration following was slightly tinged with blood.

He now, as in the first case, expressed himself relieved, and swallowed a glass of cold water, with but little effort. He was now directed a full opiate, and to drink some broth, if not too painful. In other respects to let the throat alone, and to report himself on the following day.

The next morning he appeared, walking in a dejected, stooping attitude, the head being bent to the left, to relax and favor the muscles of that side. He declared himself much worse. The pain in swallowing had recurred in the evening previous. The use of the probang had been followed by a severe chill. He had had neither drink nor food. His fauces were a little more swollen, on inspection; the uvula pendulous and flabby, and the tongue thickly coated. There was no swelling of the tonsils; efforts to swallow occasioned exquisite suffering. The mucus and saliva dribbled from his mouth, as he was unwilling to swallow them. A line extending from the spinous process of the cervical vertebræ, to the central angle of the thyroid cartilage, measured half an inch more on the left, than on the right side. There was greater tenderness, but it was more diffused. The focus of soreness was still referred to the vicinity of the cricoid cartilage.

It was therefore supposed that a spicula of bone was imbedded in the wall of the œsophagus, near the cricoid cartilage : that it was still there, because of the persistence of the symptoms ; that it was small, because it offered no resistance to the sponge probang, filling the œsophagus ; that it was imbedded, because it had not passed down, and because the expectoration was tinged with blood.

The foreign body had now been in the throat *three days*, and neither food nor sleep had been obtained by the sufferer. His pulse was 120, sharp and irritable. He confessed to the habitual use of stimulants ; and he had much less muscular power than the former patient. The imperative necessity of some measure for relief was now apparent to the patient, as well as to the surgeon : the former readily consented to an operation. He was accordingly sent (Nov. 17th) to the City Hospital; put under the influence of morphia, given subcutaneously, and the bowels cleared by a purgative injection. The operation was performed at 5, P.M., with the assistance of Dr. Thaxter, and in the presence of Drs. Homans and Swan, of the Hospital staff, and of the house-officers and a few students.

OPERATION.

The incision was made upon the *left* side. It was three and a half inches long, reaching from opposite the top of the thyroid cartilage, to the sternum; starting midway between the sterno-mastoid and the larynx, it ran down parallel to the inner edge of that muscle. The skin and platysma having been divided, the edge of the sterno-mastoid was laid bare. Next the upper belly of the omo-hyoid muscle was dissected out, and the lower boundary of the superior carotid triangle fairly exposed. The omo-hyoid was then drawn to the *inside*, or central line of the neck—the reverse of the first operation, where it was drawn outwards. This second manœuvre afforded much more room than the other method. The carotid sheath was now plainly felt, and drawn outwards. Next, beneath the edge of the sterno-thyroid and -hyoid muscles, an unusually large thyroid gland bulged into view. The veins on its surface were dilated, and the gland seemed puffed out and full. By a gentle dissection with the director, the handle of the scalpel, and an occasional touch with the edge of the knife, this gland was lifted and turned over on its isthmus, without wounding either it, or its vessels. The inferior thyroid artery, especially, was enucleated from the cellular tissue, and preserved intact. The wall of the carotid sheath was now seen in contact with the trachea. A separation was effected with the director; and the artery drawn outwards, the thyroid and trachea inwards, with blunt retractors.

So little hæmorrhage had taken place, that with a light reflected into the bottom of this deep wound, there could be plainly seen the silvery fascia and tendons of the prevertebral group of muscles, between two distinct walls, formed on the outside by the carotid sheath, on the inside by the œsophagus. No foreign body was found.

After some manipulation, as in the former case, the stomach tube was passed into the œsophagus by Dr. Thaxter, and its presence proved, both by its being felt behind the trachea, and by the gastric smell issuing from its open end. Contrary to our hopes, no foreign body could be discovered.

I now decided to open the œsophagus, and explore within. A vertical incision was made upon the stomach tube, just below the cricoid cartilage, through the wall of the œsophagus, towards its posterior aspect. This wound having been dilated, the tube was withdrawn, and the forefinger passed into the œsophagus up to the metacarpus. Searching first above, I reached as high as the top of the larynx and epiglottis, in the fauces, without finding anything. Next turning the finger downwards, I felt beneath the sternum, a long, slender, hard substance lying across the posterior wall of the œsophagus. Both ends were imbedded in the gullet, and the centre could not be lifted out. Passing down a director upon my finger I picked one end free, drew it outwards, and freeing the other end, I withdrew, instead of a bone, a common brass pin, an inch and a quarter long. It was blackened and corroded, and lay diagonally across the œsophagus, its head to the right side, just beneath the manubrium of the sternum. The head was entangled, and the point imbedded in the wall of the œsophagus.

As in the former case, the wound was left open.

Another subcutaneous injection of morphia and an enema of beef-tea were administered, and the patient placed in a room freely permeated with steam.

These precautions to keep up a warm moist atmosphere were taken because laryngitis, or spasm of the glottis, was feared. First, as a consequence of the handling and tilting of the trachea and larynx. Second, on account of the disturbance of the recurrent laryngeal nerve, which was not only liable to be stretched at the time, but to be pressed upon by subsequent effusion, when suppuration was being established. Our fears were realized, in each case, but only to a moderate degree.

9, P.M. Feels nicely; can swallow easier than before the operation; no hæmorrhage. Repeat the morphia.

Nov. 18. Has slept most of night; pulse 112. Complained of trouble in breathing, which was relieved by gargling the throat with water, and in this way removing some obstruction. Ordered absolute diet; no drink; beef-tea enemata, morphia, bathing, as in the former case.

8, P.M. Smart reaction; skin hot; swallowing more painful; wound quiet.

Nov. 19. Pulse 96; voice more clear; respiration easy; tongue and lips sore; considerable expectoration; sweats profusely. Ordered chlorate potassæ, atomized.

Nov. 20. Pulse 84; coughs a little; swallowing not so painful as yesterday; suppuration established.

6, P.M. *Three days since operation.* Complains

4

that he feels hungry all the time; allowed to drink half a tumbler of milk. A good deal came out of the wound, but did not occasion pain; his hunger was appeased.

Nov. 21. Breathing easy; voice more hoarse; drank a tumbler of milk, and lost one-fourth through the wound; wound looks sloughy. To have a glass of milk twice a day, and a half ounce of rum with each enema of beef-tea; morphia *pro re nata.*

Nov. 22. Pulse 72; slept well; retains enemata. Not much discharge from the wound, which is filled with a large gray slough, and emits a bad ordor. Can swallow easily. Took three glasses of milk and four enemata. A large part of the slough was removed in the evening.

Nov. 23. Pulse 72; slept well; swallows easily; wound cleaner and discharging; a slight cough. During the day drank *seven* tumblers of milk, losing one-fourth of it. Feels very well.

Nov. 24. *One week since operation.* Pulse 84, and strong; wound throwing off sloughs. To be dressed with a weak solution of liq. sodæ chlorinatæ.

Nov. 25. Drank *ten* glasses of milk; lost one-third.

Nov. 26. Cough continues, otherwise doing well. To try tinct. hyoscyami with the chlorat. potass.

Nov. 27. Throat feels better; cough less troublesome; wound discharges more healthy pus. Takes over three quarts of milk daily, and loses less.

Nov. 28. Looks bright and feels well; pulse 72; swallowing less painful; wound healthy.

Dec. 1. *Two weeks since operation.* Wound looking well; but little pain on swallowing, and only a small amount escapes through the wound.

Dec. 2. Wishes to omit enemata. ℞ Ferri et. potassæ tart., grs. x., ter die, in his milk.

Dec. 4. Feels that he gets enough food by the mouth; granulations of wound exuberant; sat up two hours to-day.

Dec. 5. Asks to have enemata resumed. When he drinks with his head thrown back, but very little flows from the wound; when he bends the head forward, it comes out with a jet.

Dec. 8. Wound closing rapidly; but little discharge from it; sits up four hours; consumes a large amount of milk.

Dec. 11. Has been up all day; swallows without pain.

Dec. 13. When he drinks milk, not more than two or three drops escape through the wound.

Dec. 15. *Four weeks since the operation. None of the liquids henceforward come out of the wound.* Patient sits up all day; reads and amuses himself. Continue milk and iron.

The wound was larger in the œsophagus in this case than in the first one. It closed in *four* weeks, but the other in *three* weeks. The second patient, however, *after* four weeks, made a more uninterrupted recovery than the first one.

Dec. 17. Wound has healed; to be strapped with adhesive plaster. May take soup for dinner; no pain on swallowing.

Dec. 22. Eats solid food.

Discharged well, December 29th, six weeks since operation.

SEQUELA.

The œsophageal wound was re-opened by swallowing a large morsel of beef, about seven weeks after the operation; it closed spontaneously in a fortnight. The patient is now (eighteen months since operation) well.   •

———

We think two causes can be assigned as the reason why the second case recovered so much more promptly than the first.

1st. Because constitutional irritation had not gone so far in the second. For in the first case swallowing was impossible; and that suppuration was imminent, was proved by pus appearing in the wound twelve hours after the operation.

2d. Because in the second case the wound was cleaner cut, less bruised, and the passage down to the œsophagus more direct; the omo-hyoid was drawn toward the median line, while in the other case it was drawn outwardly and snapped over the sinus and closed it on being released. We should remember, however, that the cut into the œsophagus was twice as large in the second as in the first case, but it was not bruised.

With regard to the treatment of the wound, all authors advise that it be left open, except Gross. In his Surgery he directs that the œsophagus be closed by silk sutures, cut off short, in the expectation that after union has taken place they will ulcerate through and drop into the œsophagus, as they do after injuries of the intestines. This opinion is based on no personal experience of œsophagotomy, and on no cited authorities.*

With respect to the treatment of the patient after operation, Boyer† is the first who advises absolute diet for a week, and the non-employment of elastic tubes to feed the patient through. Others, as Mr. Syme, recommend the early use of the stomach tube, or an elastic catheter passed through the nostrils to convey nourishment past the wound, without irritating it. In our cases we have followed Boyer's plan, and it seems to us the most rational. The irritation of passing tubes, however gently, is considerable; and the risk of touching granulations and breaking up adhesions is very great. For a few days the diet by the mouth should be absolute; for if we allow drink even, at an earlier period, it will infiltrate among the long muscles of the neck, and lead to suppuration. After a few days a safe sinus is established by adhesions, and drink can be taken freely and without risk. *Milk* then forms the blandest, and the most comprehensive article of food.

Boyer, after describing the lateral operation, advises us *not* to operate unless there be a projection externally.

With regard to nourishment, it is well known that

---

* It is based, however, Dr. Gross assures us, on the analogy of experiments of his own on the lower animals.

† Traité des Maladies Chirurgicales, t. vii. p. 193.

life can be supported a long while by nutritious enemata. Hennen, in his Military Surgery, gives the case of a man with a gun-shot wound of the throat, who was sustained for eighteen days with baths of milk, and with clysters. In our patients from two to three ounces of beef-extract were thrown up the rectum every six hours for many days, forming, at first, the sole nourishment. These were alternated with occasional enemata of soap and water·to clear the bowel. Water was supplied the first few days by bathing the wrists and flexures of the arm, and by wetting the lips.

CASE III.

City Hospital Records.

[Reported by Mr. L. D. Gunter, House Surgeon.]

(Service of Dr. Cheever.)—February 26th, 1868. M. M——, female, aged 60. Patient stated that in July last she swallowed about half a dozen pins, which she had removed from her dress and placed in her mouth, before taking a nap. On awaking she noticed a sharp·tickling sensation in the throat just about the level of the cricoid cartilage, accompanied by a constant desire to swallow.

These symptoms gradually increased, till at the · expiration of three weeks they became so intense that she called in medical aid. Two pins were removed from her throat, but without relief. Since then she has been gradually losing strength, and now being unable to work, owing to pain, loss of sleep and inanition, she comes to the Hospital, desirous of an operation to remove the foreign body which she is confident still remains in her throat.

The symptoms are constant pain with a burning sensation just above the level of the cricoid cartilage, and a feeling of constriction, causing an incessant desire to swallow. These are increased by swallowing either fluids or solids, especially the latter; in fact, she has been compelled to live upon liquid diet for several weeks. Complains also of similar symptoms at a point just above the level of the sternum, but they are neither constant nor severe. Soon after commencement of present trouble she suffered, more or less, from pyrosis, which seemed to increase the irritation at the lower point.

Pressure upon the œsophagus laterally at the upper point gives severe pain, principally upon the left side; at the lower point, pressure in a posterior direction gives some pain, but not half as severe as above. A sponge probang was passed down, but no stricture or obstruction of any kind could be found.

No inflammation of the pharynx, and no appearance of its recent existence. Externally nothing abnormal. No history of carcinoma, and no symptoms of hysteria. Says she had always enjoyed good health till present trouble; digestion had previously been excellent, and was so now, her only difficulty being in swallowing, and that had increased rapidly within the last few weeks. Had grown weaker, day by day, till obliged to give up work.

Feb. 27. Tested by swallowing first fluid and then solid food. The former was swallowed more readily and with less pain than the latter, but in each case an instinctive hesitation and delay was observable. There seemed to be no trouble in the larynx or trachea. Respiration good ; voice natural. She was seen in consultation by Dr. Thorndike, who concurred with Dr. Cheever, that the symptoms all pointed to a foreign body in the œsophagus, and that an operation for its removal was indicated.

Feb. 28. Symptoms increased a little, probably by examination.

OPERATION.

Etherized and placed upon the table. An incision was made, about half an inch to the left of the trachea and parallel with it, three inches in length, starting at a point opposite to the middle of the thyroid cartilage.

The subcutaneous tissues and fascia being divided with the knife, the remainder of the dissection was made with a director down to the œsophagus.

The sterno-mastoid was drawn to the outer side, the sterno-hyoid and thyro-hyoid to the inside, also the omo-hyoid. The œsophagus was reached in the superior carotid triangle. A No. 12 elastic catheter was passed down through the mouth, and the œsophagus cut through longitudinally upon it. An incision was made large enough to admit the index finger, and a thorough examination made, but no foreign body discovered. In the course of the dissection the inferior thyroid artery was cut, also another small superficial artery. But very little blood was lost, and the whole operation occupied about twenty minutes. The œsophagus was closed with six silk sutures, and the external wound by four sutures. It was impossible to prevent the escape of little bubbles of air, at first, when she swallowed; but after the wound had been closed a short time, it was entirely stopped.

Two hours after operation patient had fully recovered from ether, with but little emesis, and was feeling very comfortable. Ordered to swallow nothing. Beef tea *per rectum* every four hours.

4, P.M. No pain at seat of operation, but complains of pain in front between the breasts, and behind between the scapulæ. No hæmorrhage since operation. Respiration free; pulse 84 and good.

6, P.M. No pain except as above mentioned. Thirsty; ice to tongue. Pulse 80. Compress applied, wet with a dilute solution of carbolic acid. Laudanum, thirty-five drops, with enema of beef-tea.

9, P.M. Considerable pain in chest. Had a short nap since last seen. Wound quiet.

12, P.M. Pain still continues; ¼ gr. morphia subcutaneously. Beef tea enema every four hours if awake.

Feb. 29. Slept well after the subcutaneous injection. Very comfortable now. But very little swelling of wound. No hæmorrhage. Reaction good. Pulse

5

80. Continue beef-tea injections as before. Ice for thirst.

Evening visit. Comfortable all day. Slept considerably. Respiration free. Pulse good. Swelling extended down over clavicle, but very slight. Her only complaint is of pain in chest, front and back.

12, P.M. Cannot sleep, owing to pain; ¼ gr. morphia subcutaneously.

March 1. Slept well after subcutaneous injection. Doing well now. Wound quiet, no hæmorrhage. Swelling not increased. No appearance of pus, and but slight odor of decomposition. Slight sanious discharge, just enough to soil the compress.

Evening visit. Has had a very comfortable day. Wound looks well; swelling not increased. Discharge purulent, and a little increased, with more offensive odor. Doing well in every respect. Continue injections of beef tea during night.

March 2. About twelve last night, the pain in chest was quite severe, so much so that patient had only a short nap. Morphia sulph. ¼ gr. subcutaneously. Now quite comfortable and in good spirits. The swelling a little increased, and edges of wound slightly inflamed. The edges of the wound are apparently attached by primary adhesion. Remove sutures. Pressure upon wound did not produce any difficulty in respiration. Slight amount of pus escaped. The edges remained adherent except about an inch at the upper portion of the wound, through which the ligatures passed. Omit the injections, and give patient milk and ice water to swallow *ad libitum*.

P.M. Tested with several swallows of milk at different times during the day, but in no case did any escape from external wound. No air escaped. Says she is free from old trouble, and although her throat is swollen, yet can swallow easier than before the operation.

10.20, P.M. Hæmorrhage from upper portion of wound. Lost three or four ounces of blood before seen by house surgeon. It did not come away in jets, but a steady stream. Hæmorrhage easily controlled by slight pressure over the point of egress. The swelling above the incision has increased, undoubtedly from the clot.

Dr. Cheever was called, and, after careful examination, decided not to disturb the wound; and if the clot did not increase rapidly to let it alone, that thus it might serve as a barrier to the escape of food from the œsophagus.

March 3. Patient restless all night, but more from fear than pain. Clot not enlarged; no more hæmorrhage. In very good condition; pulse 78 and fair. Wound was opened, and a small amount of purulent fluid escaped. Rather improved since last night. Ordered sherry and milk, equal parts, at stated intervals. A little air bubbled from wound when she drank, but no milk escaped.

Evening visit. Takes her wine and milk without any trouble. None seemed to escape from external wound.

March 4. Rested very well last night. No pain, except slight aching between the shoulders. Wound looks well; a little discharge. Inject wound with dilute carbolic acid. Fœtor very offensive. No bad taste in mouth.

Evening visit. Milk came through the wound this afternoon for first time. Slough not extensive, but odor very offensive. Pulse 100, moderate strength and volume.

March 5. Sits up in bed and drinks her milk and wine. Wound looks clean and quiet. Discharge less fœtid, but about the same in amount. Milk escapes through the wound.

March 6. Doing well. Wound cleaner.

March 7. Discharge a little more profuse; probably half a drachm of pus escaped upon pressure from the deep portion of the wound. Complains of feeling hungry. Ordered milk and wine, beef essence, and steak to chew. Sat up in chair a while to-day. She thinks about half of what she swallows escapes through the wound.

March 8. Makes no complaint of pain. Wound looking remarkably well. Removed all the ligatures, and a large mass of fascia from the deep portion of the wound. The injection occasionally passes into the œsophagus, and can be distinctly tasted, although it does not come up into the mouth. No hæmorrhage; discharge much less; not over half a drachm at a dressing. Wound perfectly healthy; granulations sprouting out as far in the wound as visible. The edges of the wound have lost their former red and irritable appearance. Apparently swallows with greater ease than before the operation. A slight tenderness and pain in side of neck just behind the wound. Ordered fomentation of hops. Diet—milk, wine, and steak to chew.

March 9. The difficulty caused by the escape of milk and wine through the external wound, and the consequent inanition, has been entirely overcome by pressing upon the wound with a folded compress. Takes freely of milk and eggs beaten together. Feeling better and stronger. Wound clean and granulating. No pain.

March 10. Sat up a great part of the day, and walked about the ward. Strength improving. Wound clean, and no fœtor. *Twelfth day since the operation.*

March 12. Wound clean, and healing rapidly.

March 14. Wound nearly one third closed. Chicken broth, eggs and milk ; beef steak to chew.

March 16. Feeling well; appetite good; and wound progressing rapidly.

March 17. No discharge from wound, which appears about half as large as at first. Feels none of the former symptoms in her throat, and swallows bread and milk without difficulty. Very slight pressure from without inwards, upon the sterno-mastoid muscle, prevents any escape of liquids from the wound.

Evening visit. Wound closed, except a small portion, the size of a No. 2 catheter. When swallowing her food in small quantities at a time, none escapes from the external wound ; when increased amounts are swallowed, a little escapes, but not one

tenth of the former amount. Superficially the wound
is open one half its length. Assists in doing the ward
work.

March 18. *Nineteen days after operation.* Dr.
Cheever tested her ability to swallow. She took
large and small swallows alternately, but not a drop
escaped through the external wound. Diet—bread
and milk, upon which she is gaining flesh rapidly.

March 20. Nothing escaped through wound since
previous date. But little difficulty in swallowing,
especially fluids. Wound is less than an inch in
length, and superficial.

March 22. Wound healing rapidly, is but a simple
superficial ulcer. Strap.

March 24. Swallows liquids and solids with little
difficulty.

March 26. Improving. Wound nearly healed.

March 28. Only a small, superficial granulating
surface. Milk, bread and butter and steak swallowed
without difficulty or hesitation. Entirely free from
irritation and pain, which were so troublesome at
time of entrance. Respiration good; voice natural.
No swelling about seat of operation, which leaves
but a small cicatrix. Discharged well. Requested
to report at Hospital once a week, for a time.

REMARKS.

At the time of the operation, two things were
noticed :—An indurated spot outside the œsophagus
and near the cricoid cartilage ; and a valvular condition
of the mucous membrane of the œsophagus. Whether

the first was the former resting-place of a pin, which had come through the walls of the œsophagus ; or whether the last could explain the difficulty in swallowing, are questions unsolved. The operation, at any rate, did no harm. The patient left the Hospital in four weeks—encouraged, and free from suffering.

SEQUELA.

After being closed six weeks, the œsophageal wound re-opened, after hard labor and exposure in washing. The patient re-entered the Hospital. The fistula closed in a few weeks. She is now, four months since the operation, perfectly well.

Sept. Six months since operation; the fistula has re-opened.

HISTORY.

WE now pass to the History of the operation, embracing the opinions and experience of high surgical authorities; an analysis of the different varieties of foreign bodies by M. Hévin ; the memoir on œsophagotomy of Guattani, the earliest systematic writer on the operation ; a detailed account of the operations of Bégin, Martini, Delarocherie, Arnott, Cock and Demarquay ; an analysis of cases and the results of experiments on the lower animals by M. le

Docteur Créquy; and finally, such brief general rules as we can deduce from these authorities for our guidance.

We will examine first the opinions of Fergusson, Nélaton, Syme and Velpeau. Sir William Fergusson, in a very brief allusion to the operation in his Surgery, cautions his readers, that however simple œsophagotomy may appear on the dead subject, it is attended with much danger on the living, and we should proceed with great caution. He mentions three cases of foreign bodies in the œsophagus.

I. An epileptic swallowed his false teeth. Four years and a half afterwards he died from bronchitis. The teeth were found at the lower part of the pharynx. The attending-surgeon had not operated, for fear of causing fatal laceration.

II. The metallic setting of some false teeth was swallowed, and caused death by perforation of the arch of the aorta.

III. The only case Sir William saw himself. A little girl swallowed a small padlock. It could not be felt or seen; she was not believed in her statement. After suffering some days, coughing and vomiting came on, the foreign body was thrown up into sight, and withdrawn through the mouth.

Certainly a moral might be drawn here, rather in favor of the operation than against it.

Nélaton. "Elemens de Pathologie Chirurgicale."

Œsophagotomy. "We must not conceal from ourselves the dangers inherent in this operation, which is, certainly, one of the gravest and most difficult in surgery—we must, therefore, advise it only in cases where other means have failed.

" But in view of the accidents which arise from the impaction of a foreign body in the œsophagus, such as asphyxia and death; or, consecutively, from its continued presence, such as perforation of the œsophagus, of the aorta, of the carotid, et cetera, the surgeon ought not to remain inactive and a passive spectator of consequences which, in a great number of cases, lead to the death of the patient; and which, in other cases, cause him to undergo dangers, often greater than those which are peculiar to the operation itself."   •

" Should we operate only when the foreign body can be felt from the outside? Although this is a favorable condition, we do not think the absence of it a contra-indication to the operation. We think, then, that the operation should be performed, when the foreign body cannot be withdrawn through the mouth, and when it cannot, or ought not to, be pushed down into the stomach."

According to M. Bégin, when a sharp foreign body enters the œsophagus, the muscular fibres, irritated into excessive action by the pricking of the mucous coat, contract spasmodically above and below it, and hold it inclosed in a sort of sac, whence the efforts of neither swallowing nor vomiting extricate it.

Nélaton advises to operate on the median line.

1st. By making the incision for tracheotomy.

2d. Separating the sterno-hyoid muscles.

6

3d. Passing a double ligature on either side of the thyroid gland.

4th. Tying, and then dividing the isthmus of the thyroid gland.

5th. Turning over the left lobe of the thyroid gland and finding the œsophagus beneath and to the left of the trachea.

The objection to the median operation would seem to be that we fall upon the front of the œsophagus, where the recurrent laryngeal nerve runs, between the gullet and the trachea, and thus endanger the nerve or its filaments more than by the lateral operation, in which we attack the œsophagus at the side and behind. And the median cannot be more free from hæmorrhage than the lateral operation, which can be conducted (vide Case II.) without dividing any vessels large enough to require ligature.

M. Nélaton continues: "As a *resumé* of diagnosis; the history of the case, an examination outside the neck, inspection of the fauces, and catheterism of the œsophagus, suffice, usually, to make it clear.

"The *prognosis* is easily deduced from the facts previously stated. Speaking generally, it may be considered a grave accident when a foreign body retains its position in the œsophagus; for operations for its extraction are far from being innocent, and if these attempts fail, the retention of the foreign body subjects the patient to most serious risks."

These risks are primary, and secondary.

Primary, suffocation and hæmorrhage.

Secondary, inflammation, suppuration, perforation, caries, et cetera.

" If the foreign body remain, and lead to suppuration, it may be vomited up, but most frequently retains its position where it was arrested, and the suppuration becomes so abundant, with emaciation, fever and purulent expectoration, as to give the appearances of phthisis. It may happen that the abscess formed in the walls of the œsophagus, after perforating them, produces purulent deposits in the neighboring parts, which carry with them the foreign body, and may cause deep-seated disorganization in the neighboring tissues. Some of these purulent collections open and discharge at the side of the neck ; others at more distant spots, and still others penetrate the chest, and discharge pus and alimentary matters into the pleural cavity."

It will be remembered as an anatomical point, that the superior thoracic fascia arches over the œsophagus at the upper border of the thorax, and leaves a free passage for pus to gravitate behind the œsophagus into the posterior mediastinum.

" Among the secondary dangers," continues M. Nélaton, " there are some, which though less frequent than inflammation, are none the less to be feared ; such are perforations of blood-vessels or the trachea, in consequence of adhesions established between these organs and the œsophagus. Thus the aorta, the carotid, the right subclavian, the vena cava, the vena azygos, have all been perforated by foreign bodies through the œsophagus, with fatal hæmorrhage as the result. Dupuytren had a patient who coughed up his food, and died in consequence of the false passage. Andral reports a similar case.

" Finally, an organic disease of the vertebræ may be set up by the presence of a foreign body in the œsophagus. A child of twenty-two months swallowed a little flat, triangular bone. Two months later it died, of profound marasmus, and the posterior wall of the pharynx was found pierced, and caries of the vertebræ opposite. Analogous facts have been observed in connection with retro-pharyngeal abscess."

" It may happen," says M. Bégin (op. cit.), " that the foreign body cannot be withdrawn by the mouth, and that its presence does not immediately threaten life. Some advise, then, to wait; the foreign body may be expelled by suppuration, may soften, decay, or fall into the stomach. This plan of temporizing offers real dangers; and an experience of twenty unfortunate results, warrants us in prohibiting it."

Mr. Syme says, " The object of the operation of œsophagotomy is to make an opening into the œsophagus to extract a foreign body that cannot be removed otherwise; but such a circumstance happens so rarely, that there are few instances of a surgeon being called upon to perform the operation in the whole course of his practice. Upon two occasions I have cut into the throat for the removal of a piece of bone, which had passed through the coats of the œsophagus and caused deep-seated suppuration.

" In the first case the bone had been in six days, and I found the patient with a very anxious expression of countenance, and slight fulness of the neck, which was not discolored, but tender on pressure. She had, also, fits of dyspnœa. Not being able to reach the foreign body, and fearing it would cause suppuration, if it had not already done so, I decided to operate by œsophagotomy. In the second case the bone had been in the throat sixteen days; the voice was husky and breathing difficult; a deep-seated abscess was opened, and the foreign body found and removed.

" In a third case, a copper coin had been impacted in the œsophagus, more than two months, opposite the sternum. All these cases recovered.

" Cutting through the coats of a sound œsophagus, for the removal of a foreign body, is an operation that can hardly ever be required, since, before it is warranted, there will almost necessarily have elapsed such a period of time, as must allow ulceration and suppuration to be induced by the intruder. Any hard substance so situated should be regarded with much apprehension, since, if urged in the wrong direction by probangs, it must expose the patient to extreme danger, from the formation of matter behind the larynx, or ulceration into the air-passages, or vessels."

Now it is precisely from these perils, that we claim that an *early* operation saves the patient.

Velpeau says of œsophagotomy, " Though this operation may not have been formally proposed by any person before Verduc and Guattani, we cannot avoid seeing that the idea of it is found in more ancient authors. The incision into an abscess containing a small bone, which had escaped from the œsophagus, was done by Arculanus and Plater; fish bones were extracted in the same way, by Honlier and Glandorp. But wounds of the œsophagus had been considered so dangerous that practitioners hesitated to operate.

"Œsophagotomy was done for the first time by Goursault, in 1738; and next by Roland, in 1819. Since this operation has taken rank among the systematized ones of surgery, it has received various improvements. But, according to B. Bell, there can be no settled plan for the incisions, as that depends upon the position of the foreign body. Bell knew that, with precaution, the recurrent nerve could be avoided. As to the after treatment, a gum elastic tube must be introduced through the nares or mouth, beyond the seat of operation, and not be withdrawn for three or four days, in order that aliments and drinks may not interfere with the agglutination of the wound of the œsophagus, or escape into the tissues of the infra-hyoid region."

He adds, "If œsophagotomy, performed on the thirtieth day by Mr. Arnott, proved fatal, yet two patients, operated on at the Val de Grace, by M. Bégin, one on the eleventh, the other on the sixth day, were perfectly restored."

The verdict of M. Velpeau would therefore seem to be against delay in operating.

HEVIN.

*Précis d'Observations sur les Corps étrangers arrêtés dans l'Œsophage.**

IN a very exhaustive article on the above subject the author divides foreign bodies into the following classes.

1st. Those which can be pushed down into the stomach.

2d. Those which ought to be withdrawn by the mouth.

3d. Those which should be withdrawn, but which must be pushed down.

4th. Those which can neither be withdrawn, nor pass through the natural passages.

5th. Those which ought to be removed by incision.

In the first class are pieces of money, morsels of food, and bodies, generally, without great bulk or inequalities. The chief danger from this class is suffocation.

In the second class, are needles, pins, fragments of glass, blades of iron, nuts, and generally, angular and hard bodies. The dangers from this class are abscess and perforation.

In the third class, are those of the second which have passed beyond our reach, and for which no other natural exit remains but by pushing them down. And although the latter alternative is full of risks, it should be taken rather than to employ reiterated at-

---

* Mémoires de l'Academie Royale de Chirurgie, tome i., 1819.

tempts to remove them by the mouth ; such repeated efforts at withdrawal augment the danger.

Fabricius Hildanus relates the case of a young man who swallowed a small bone, and tried in vain to reject it by vomiting. Fabricius endeavored to touch the foreign body, first with a catheter, then with a sponge, but could recognize only a sort of strictured portion of the œsophagus, and the seat of a dull pain.

Judging that renewed attempts would only irritate the parts and occasion greater danger, he prescribed remedies to allay the inflammation. But the young man being impatient addressed himself to a village surgeon, who, by the introduction of various instruments into the œsophagus, irritated it to such a degree, that swallowing became impossible, and respiration impeded from the swelling and tension which ensued. Fabricius· being recalled, employed, in vain, remedies to relieve him. At the autopsy the pharynx and œsophagus were found.sphacelated, and the lungs inflamed, but the bone could not be found.

The bones most liable to become engaged in the œsophagus are the sharp spines of the fish's vertebræ. M. Hévin gives many instances, too long to quote here—and included in the fourth class. The last class includes cases of trachcotomy, œsophagotomy, and gastrotomy.

GUATTANI.

The earliest distinct account of the operation is an
Essay on Œsophogatomy, by M. Guattani, in the
Memoirs of the Royal Academy of Surgery, Vol. 3,
1819. This author gives the result of numerous
experimental operations on animals, describes at
length, and insists upon the operation in the human
being, and narrates the following melancholy case.

" The patient, a man forty years of age, threw up
into the air a boiled chestnut, and catching it in his
mouth, it became arrested in his œsophagus. He
complained of being unable to swallow, and was sent
to the Hospital of Saint-Esprit, in Rome, of which I
was in charge. Having examined and questioned
the patient, the fact was doubted, both because he
was intoxicated, and because he breathed and spoke
without difficulty ; and also, because he had vomited.
Moreover, there was no tumor visible from without ;
yet, since he felt pain on the left side when the larynx
was pressed, we suspected that the chestnut might be
engaged in the œsophagus. The usual remedies
were tried. The bougie was used with difficulty,
because the lower jaw was in convulsive action, and
the finger could hardly be introduced. The countenance was congested, the tongue moist, the pulse
frequent, he complained of great inward heat, and
showed signs of delirium. Since he could not swallow, nutritious enemata were given ; he desired
nothing but cold water and ice. The sixth day, notwithstanding frequent bleedings, he had nose-bleed.

The eighth day his respiration began to be impeded; fever continued, with a feeble pulse; he coughed continually, and the expectoration was purulent. The tenth day he was so prostrated, that he was thought to be near his end, yet he swallowed a little wine and broth. A peculiar noise was heard when he swallowed. The fifteenth day he had renewed epistaxis. The nineteenth day he died."

" The autopsy revealed an adhesion of the œsophagus to the larynx and trachea, with a sphacelated perforation between the two tubes; an abscess, containing the chestnut, enclosed in a sort of pouch of the œsophagus, and situated just *below and to the left of the cricoid cartilage.* The abscess pressed on the trachea, and burst as soon as it was touched."

" It is assuredly in such a case," continues M. Guattani, " that an operation is absolutely indicated. Verduc well said that it could be hazarded; he acknowledges that the operation is difficult, but he adds, ' it is better to undertake it than to see the patient die.' I am entirely of his opinion; the anatomical structure of the parts to be operated on assures us that œsophagotomy is practicable."

M. Guattani then goes on to describe the mode of operating, preferring the central incision, as in tracheotomy, between the sterno-hyoid muscles, and thence down by the side of the trachea to the œsophagus. He advises the stopping of venous hæmorrhage by pressure, and instructs how to avoid the recurrent laryngeal nerve. He disengages and lifts over the thyroid gland. He urges an early operation, if we operate. He gives very little nourishment by the mouth at first, but mostly by the *rectum*.

Here follow three experiments.

1st. " The 9th of March, 1747, I did œsophago-tomy on a dog. I united the lips of the wound by two sutures. I then applied compresses and uniting bandages. The next day, I made the dog swallow some milk by force; on the fourth day he began to eat soup. In the evening I removed the dressing ; the cicatrix was well formed, excepting a little sup-puration in the sutures ; there was a species of tumor as large as a pigeon's egg. On pressing it, nothing escaped. The dog being entirely well after a week, I thought best to kill and examine him. I found that the cicatrix of the wound of the œsophagus was formed by adhesions to the neighboring parts, as happens generally in the intestines. By means of these adhesions a little pouch had formed, in which liquid was found. I suspected that this defect had been caused by the food, which I had forced the dog to swallow too soon."

2d. " The 15th of April, 1747, I operated on another dog. I approximated exactly the lips of the wound, and applied on either side graduated com-presses, and the uniting bandage. No food was forced upon him ; but the day after the operation he ate some soup, set before him. He was entirely well in nine days. I dissected the dog to examine the cicatrix of the cut in the œsophagus, which had been made obliquely ;. *it was perfectly united, and without any adhesion to the neighboring parts.*"

3d. "At the Hospital of *La Charité*, in Paris, in presence of M. Faget, Surgeon in Chief, I operated on a human subject, which was œdematous, and which had the thyroid gland so swollen that.it covered the œsophagus. Notwithstanding all these obstacles, which seemed to be the only ones which could be found in this operation, M. Faget saw that it was easy to open the œsophagus, without danger."

Richerand says, "However exact may be the anatomical knowledge of the operation, the risks are too great to render it practicable."

Delpech says, " A foreign body producing alarming inflammation in the œsophagus, ought to be extracted by an operation. Experience has sufficiently proved the possibility, and the little danger of such an operation."

We should remember, when considering the rapid closure of the œsophageal wounds just recorded, that they were made on the lower animals, and that their gullets had not been irritated by foreign bodies.

We adduce the following as a typical case of a small foreign body impacted in the walls of the gullet, and of the usual termination of such a case, if the fragment remains.

Case of non-removal of foreign body, in œsophagus, followed by death; by Thos. Cock, M.D., 1809.*

" Mrs. H., æt. 73, eating without teeth, swallowed a piece of meat without chewing, and complained that it was lodged in her throat. Various domestic

---

* Eve's Cases in Surgery.

attempts to remove it failed ; a physician gave emetics, without effect. She could still swallow liquids. A surgeon next introduced a probang, with much pain, repeatedly, but without relief. The surgeon's fingers failed to reach the obstruction. The probang passed to the stomach. She could now swallow no fluids. She continued in this situation thirteen days, without nourishment, and in great suffering. She still insisted that something was lodged in her throat. Dr. Cock saw her the day she died, and proposed to operate, but it proved to be too late.

" The autopsy revealed a large abscess communicating with the œsôphagus, and containing a fragment of jagged bone, an inch and a half long."

Thus proving the entire inadequacy of all means taken to remove it; and that early opening the œsophagus from without, might have afforded relief.

Also the following CASE.

Retro-pharyngeal abscess, mistaken for œdema of the glottis ; laryngotomy ; death.*

" The patient, a man of forty years and intemperate habits, was admitted to the Hospital, Sept. 27, 1837, under the care of M. Ballot (Voyez Archives Générales). The pharynx was red and dry; deglutition difficult, and he was unable to speak aloud. He complained of a foreign body in the throat. The finger felt a tense, elastic swelling, on a level with the larynx. He was bled eighteen ounces, and treated

---

* American Journal of the Medical Sciences. April, 1842.

antiphlogistically. Next day, bled again eighteen ounces ; thirty leeches to throat, and blister to nape of neck.

" Laryngotomy, Oct. 8th, to prevent suffocation. During the night the tube became displaced, and he died asphyxiated.

" Autopsy. No tumefaction of glottis, but its orifice closed by a fluctuating swelling, the size of a hazel-nut, which projected over it. This was a collection of healthy pus between the vertebral column and the œsophagus."

" In any similar case, tracheotomy should be preferred to laryngotomy."

ARNOTT.

Mr. Arnott, at the date at which he writes, namely, 1833, can find no record of the operation having been done in England.*

The following extracts comprise the essential parts of his case.

" December 22d, 1833, I was called to see a boy, two years and a quarter old, who had swallowed, six days previously, a portion of mutton bone. Nothing but fluids had been swallowed since. On introducing the finger, to its utmost extent, deep below the glottis, on the right side, a piece of bone could just be touched, projecting upwards. Forceps, hooks, emet-

---

* Case of Œsophagotomy, with Remarks, by James M. Arnott, Surgeon to the Middlesex Hospital. Medico-Chirurgical Transactions, Vol. 18.

ics, and so forth, were tried in vain. There were no urgent symptoms, and some hope was entertained that the bone might become loosened and be expelled or swallowed. For a fortnight there was no suffering, but emaciation. On the 16th of January he was again brought to the Hospital, and I again felt, and tried to extract the bone. After consultation, and some delays due to the parents, I determined to operate, particularly as the child's breathing was becoming occasionally oppressed. The bone formed no prominence, and could not be felt externally. As it was fixed at the junction of the pharynx and œsophagus, I decided to cut behind the cricoid cartilage. The operation was done on the 21st of January, four weeks after the accident. The incision was made midway between the sterno-mastoid and trachea ; the deeper dissection made by the handle of the scalpel, and the œsophagus was incised on a male catheter introduced by the mouth.

" The spinous process of one of the lower dorsal vertebræ of a sheep was withdrawn. The child bore the operation well. It was fed through an elastic catheter. The breathing became more labored, and it died fifty-six hours after the operation.

" On examination of the body, there were two points of superficial ulceration, or rather abrasion, on opposite sides of the œsophagus. There was no suppuration between the pharynx and œsophagus, and cervical vertebræ. The larynx was normal ; the lower part of the trachea and bronchi were inflamed ; the right lung was hepatized ; the left lung less so."

Mr. Arnott thus sums up his opinion of the operation.

" The occasional necessity for the operation of œsophagotomy has been recognized by most systematic writers on surgery, and they have adduced certain circumstances as conditions justifying its performance. They have stated that when a foreign body cannot be extracted, or conveyed into the stomach—when it impedes deglutition, and by pressure on the trachea threatens suffocation—*if it projects externally*, the operation of œsophagotomy ought to be performed. Now in this case which has been related, deglutition was *not* totally interrupted; there was *no* apprehension of suffocation from pressure on the trachea; the body did *not* project externally, and yet the operation was requisite.

" That extraneous substances have occasionally remained for some time in the œsophagus, as in other mucous canals, the trachea, bronchi and urethra, without producing threatening symptoms, or immediate injury, and have afterwards been expelled, or made their way out, is well known; but experience has shown that fatal consequences, from ulceration of the larynx, trachea, and carotid artery—all of which have taken place—may likewise ensue. The issue of the present case proves that an additional sequence is to be feared in the form of pneumonia.

" The rule of practice, then, ought to be, when a solid substance, though only of moderate size and irregular shape, has become fixed at the commencement of the œsophagus, or low in the pharynx, and has resisted a fair trial for its extraction or displacement, that its removal should at once be effected by incision, although no urgent symptoms may be present.

" I insist the more upon this point, because I have reason· to believe that the operation has been in several instances omitted, when it ought to have been performed, chiefly from its supposed danger and difficulty, but also from a reliance upon the powers of nature. I have particularized the low situation of the substance, as a reason against unnecessary delay, because great difficulty is experienced in seizing it when so placed, and the attempts will, I believe, usually fail.

" By most writers œsophagotomy has been represented as only warranted when the foreign body projects externally, which they seem to consider requisite as a clear indication of its presence, and as a guide for the division of the soft parts. This projection *may facilitate* the operation, but is *not* essential to its safe performance. As long as we are certain of the presence of the extraneous substance, the operation may and ought to be undertaken, even though it not only does *not* project, but cannot be felt externally."

8

COCK.

In connection with the Records of a Case of Pharyngotomy by Mr. Edward Cock,* we find the following remarks : " The operation appears to have been one of great rarity, and I can find but very few cases recorded in which the pharynx and œsophagus have been opened for the purpose of extracting a foreign body. A vast number of cases are on record of foreign bodies retained in the gullet. In some instances they have been extracted through the mouth, in others they have been pushed into the stomach ; or the means employed having failed to produce either of these solutions of the difficulty, they have been left until some favorable process or fortunate effort of nature has got rid of the mischief ; or else, as has frequently occurred, the patient has perished either from the retention of the body in the gullet or from the violence which had been used to effect its dislodgement.

" Directions for exposing and opening the gullet may be found in several books of operative surgery ; but, generally speaking, they are vague and speculative in their character, and evidently not founded on the results of practical experience.

" In the year 1832, M. Bégin had the satisfaction of performing the operation of œsophagotomy in two cases, with perfect success. A soldier had swallowed a beef-bone, which stuck in the œsophagus, at the

---

* Guy's Hospital Reports. Third Series, vol. iv., 1858.

lower part of the neck. Various attempts were made
to dislodge it without success, and on the twelfth day
after the accident, the operation was performed.
Much difficulty was experienced, but the bone was
eventually extracted, and the patient speedily recovered. In the second case, M. Bégin . removed
from the lower part of the cervical portion of the
œsophagus a large fragment of bone swallowed eight
days previously. The patient recovered.

"In 1845, Dr. Martini incised the neck for the
purpose of extracting a portion of bone, which had
been swallowed four days previously. The foreign
body could be felt from the exterior projecting above
the clavicle, and the incision was made on that spot.
It would seem, however, that the patient anticipated
the intentions of the surgeon, and swallowed the bone
before it could be seized and extracted. Death from
collapse took place two days after. The pharynx
was found in a gangrenous condition, and the stomach
and duodenum were inflamed. The bone had passed
into the rectum. There can, I think, be no doubt,
that in this case the fatal result was brought about,
not by the operation, but by the severe and somewhat
extraordinary means which were previously used to
dislodge the bone. When we read that the man was
repeatedly bled, that sixty separate attempts at dislodgement were made with levers and forceps, that
enemata of belladonna were employed, that, finally,
tartar emetic was injected into the veins, followed up
by clysters of vinegar and opium to counteract its
effects ; moreover that during this ordeal the patient

was unable to swallow even a drop of water, it is not surprising that he finally succumbed.

" In 1845, also, M. Delarocherie, Professor of Clinical Surgery at Liege, removed a large portion of bone from the gullet of a man by œsophagotomy. The operation was performed the eighth day after the accident, and not until the patient's life had been nearly compromised by the repeated and strenuous efforts made to dislodge the foreign body, giving rise to profuse bleeding and severe injury to the gullet, which seems to have become ulcerated through its walls previous to the operation. The wound sloughed; but the patient had recovered by the twenty-sixth day."

Mr. Cock's CASE.—" Mr. T. G——, having been in the habit of wearing a false central incisor tooth, fixed in a gold plate, which extended some distance on either side, while asleep, January 17, 1856, it became detached, and he swallowed it. There could be no doubt that the foreign body had lodged in the cervical portion of the gullet, but its exact situation was not clearly indicated. The irritation, pain and tenderness were referred to the top of the œsophagus, just below the larynx ; but no projection could be detected outside. The patient swallowed fluids with great difficulty and in small quantities. The breathing was not impeded, but he had an irritating laryngeal cough. A bougie passed down found a total obstruction about the lower edge of the larynx. A pair of forceps detected the plate, but it could not be grasped or moved.

" The next day it appeared very doubtful whether any fluid, which he took into his mouth, went into the œsophagus. Attempts were made with several instruments to grasp the plate, but they proved abortive. A number-five flexible catheter was passed down beyond the obstruction, and through this an emetic was conveyed into the stomach. It failed to excite the slightest nausea. A mode of administering nourishment had, however, been obtained, and it was decided to wait a little longer. On the 19th, another fruitless attempt was made to dislodge the foreign body. He could not swallow a drop. On the 21st, some operation was imperative. Œsophagotomy was performed on this, the fourth day since the accident. The operation was the lateral one. The omo-hyoid muscle was divided. Several arteries were tied. The foreign body was found behind the cricoid carti-lage, and extracted.

" The patient was fed through a catheter, and afterwards through a stomach-tube, and not allowed to swallow. At the end of three weeks, he swallowed; at the end of four weeks, the wound was closed."

It will be noticed that this period was about the same as in our three cases, where no tubes were em-ployed, and the patients allowed to swallow freely after three or four days, the liquids passing constantly through the wound.

" The operation, however, had produced a very decided alteration in the voice, which had previously been clear and strong—when he left the hospital he spoke in a hoarse whisper. Indeed, there was an

almost entire loss of those tones which appear to depend on the tension of the vocal cords, and as there
was no reason to suppose that the mechanism of the
larynx had suffered from the temporary lodgement of
the foreign body, it seems fair to infer that the filaments of the recurrent laryngeal nerve, which supply
the arytænoid muscles, had been partially or wholly
divided in the operation.

" Two years later, the patient was well, and the
voice strong, but it had undergone a decided trans-
formation.   Formerly tenor, it was now bass.

" Much valuable instruction may be gained from
the monographs of Bégin and Arnott, whose decisive
views and opinions on the subject, founded on practical experience, may well supersede the more timid
and vacillating policy advocated in many books of
surgery.   Œsophagotomy has been too rarely performed to allow us to speak with absolute certainty
as to the dangers which may be incident to it as an
operation, or its positive success as a remedial measure.
There can, however, be no doubt that the fatal result
in the two cases out of seven recorded, depended, not
on the operation, but in the one on the delay and the
presence of severe pulmonary mischief, and in the
other on the unmerciful palliative treatment which
preceded the use of the knife.   It is somewhat
singular, that in all the successful cases the precise
locality of the foreign body was neither marked by
any external prominence, nor indicated by any accu-
rate reference of pain to one particular spot.   Indeed
we can hardly expect that either of these conditions

should ever occur, although their presence has been strongly insisted upon as necessary guides in the performance of the operation, and, indeed, as a *sine qua non* as to the propriety of undertaking it.

" However slender our experience may be, the feasibility of the operation is established by the facts, that it is quite possible to make a long deep incision along the side of the neck, and to continue the dissection down to the bodies of the vertebræ, without compromising any vitally important organ. That a considerable longitudinal incision can be made into the gullet, whether pharynx or œsophagus, without present mischief or future detriment. That experience shows how much more likely the gullet is to suffer from the retention of the foreign body or the means used for its extraction, than from the knife of the surgeon.

" I am inclined to believe that a foreign body might be extracted from any portion of the cervical gullet, from the pharynx to the sternum. We might even extract it from the thoracic portion of the œsophagus, if it could be reached with forceps.

" The most accessible part of the gullet is, doubtless, the two first inches of the œsophagus, included between the upper and lower thyroid arteries. Below this point, as we approach the upper opening of the chest, the difficulties of the operation increase in consequence of the closer proximity of the carotid sheath to the trachea and œsophagus."

CREQUY. *

" The opportunity having been afforded us, while serving as *Interne* to M. Demarquay, to observe two cases of this kind, we have looked to see if other analogous cases could be found, and if any practical deductions could be derived from them.

" CASE I. A *sou* arrested in the upper part of the œsophagus, leading to a perforation, a retro-pharyngeal abscess, and finally perforation of the pleural cavity and death.

" The 26th of June, 1855, a little boy of five years of age, while playing with a sou, which he threw up in the air and caught in his mouth, accidentally swallowed it, and it became arrested in his œsophagus. The child suffered but little at the time of the accident. A physician, who was called, administered an emetic, which acted; but from this moment severe pain was felt, a sense of oppression, and inability to swallow. M. Demarquay saw the child the 27th, and passed down Graefe's sound. The instrument touched the foreign body, but, after several attempts, he could not dislodge it. The child now felt better, and resumed its play; but in four hours complained of severe pain in the neck; he began to cry, and the parents perceived that the head and neck began to swell.

* " Observations de Corps Etrangers arrêtés dans l'Œsophage. Œsophagotomie ; opportunité de cette operation, par le Docteur Créquy, ancien interne des hôpitaux."
Gazette Hebdomadaire de Médecine et de Chirurgie. Nov. 1861.

" On the 29th, the child entered the *Maison Municipale de Santé*, in the service of M. Demarquay.

" He now lay upon his back ; the cheeks, the neck, and the chest below the clavicles, were swollen, and mostly on the right side. A fine, emphysematous crackling was felt on pressure over these parts.

" The swallowing of liquids was difficult, and of solids, impossible ; the drinks were regurgitated with a large quantity of saliva. The respiration was impeded, and accompanied with a gurgling sound. Fever and anxiety, increasing.

" On the 3d of July, four days after his entrance, he sank and died. •

" The autopsy showed the anterior and posterior mediastinum infiltrated with pus and air. The trachea inflamed. At the junction of the œsophagus with the pharynx, on the right side, a perforation about one and a half centimetres wide, communicating with a retropharyngeal abscess. This extended from the basilar process of the occiput, down to the fifth dorsal vertebra. The pharynx and œsophagus were completely separated. The carotid, the jugular, and the pneumogastric were denuded by pus. The *sou* lay at the bottom of the abscess, penetrating the right pleura, by a perforation. Through this the drinks taken had passed into the pleural cavity. There was another opening from the posterior mediastinum into the pleura."

9

CASE II. " A *franc* piece caught in the upper part
of the œsophagus. Futile attempts at extraction;
œsophagotomy; retro-pharyngeal abscess; death.

" The 26th of July, 1854, a little girl of three
years and a half swallowed a franc piece, which
several people saw, at the bottom of her throat. One
of them pushed it down farther, and it could not be
seen or touched afterwards. M. Demarquay passed
Graefe's sound. It was arrested at the lower part of
the neck by an obstacle which it passed with difficulty;
but in withdrawing it, the hook glided over the
foreign body, and all attempts at its extraction failed.

" These attempts were renewed three times within
as many days. They caused some bleeding, and great
suffering.

" On the 28th of July, she was admitted to the
Maison Municipale de Santé.

" On the 5th of August, the child's condition was
constantly growing worse; the oppression was considerable, deglutition very difficult, and the saliva was
mixed with pus.

" Œsophogatomy was performed on this, the 10th
day since the accident. The incision was made on
the edge of the sterno-mastoid muscle, and the œsophagus reached between the sterno-thyroid, the
trachea, and the carotid sheath. Very few, and small
vessels were tied. No important parts were injured,
as proved by the autopsy. The foreign body was
found and extracted.

" August 6th. The child suffers but little; the
respiration is easy, pulse 120. The patient is much

enfeebled from inanition, having taken no solid food for eight days. The drinks swallowed now run out of the œsophageal opening. M. Demarquay endeavored to nourish the child by a tube passed down the œsophagus. These efforts were vain, as efforts at regurgitation rejected the drinks taken. The œsophageal opening was gaping and *fetid*, a peculiarity common to all wounds of the intestinal canal, wherever made.

"August 7th. Pulse more frequent; a gurgling sound heard at the lower part of the neck. This sound, resembling that heard in the former case, makes the surgeon fear a mediastinal abscess.

"August 8th. The child died.

"Autopsy. Beside the larynx a small, circumscribed abscess, which has no connection with any other parts. Opposite the. upper border of the larynx, on the left side, a small perforation of the posterior wall of the œsophagus, with dark, gangrenous edges.

"This opening, caused, probably, by the pressure of the foreign body, communicates with a retropharyngeal abscess, which after having dissected up the posterior wall of the pharynx, opened into the right pleura. Below the larynx the posterior wall of the œsophagus has several perforations : one, on the left, made in the efforts at extraction during life ; two others, above and below the right bronchus, communicate with the large abscess. No changes in the other organs.

"In these two cases the pathological changes, due

to identical causes, presented a remarkable analogy. The piece of money caught at the commencement of the œsophagus, set up an irritation in the walls of the canal, which communicated itself to the neighboring cellular tissue ; pus formed rapidly in the latter, and there resulted a large purulent collection, which opened into the pleura.   In the first case, the purulent collection was evidently preceded by a perforation of the alimentary canal ; since this perforation took place the second day after the accident, as shown by the symptoms.   In the second case, nothing indicates that it was so ; suppuration must have taken place in the cellular tissue around the œsophagus, before the wall of the canal was perforated.

    " Facts which we have sought for in various authors do not enable us to decide this question ; yet, if we may judge by what takes place in other canals, we may believe that for an abscess to be formed it is not necessary that the wall of the œsophagus should be first perforated by a foreign body.   It is thus that fæcal accumulations, the stones and seeds of fruits, and other similar bodies, arrested in the large intestines, give rise to abscess of the iliac fossa, without producing a rupture of the walls of the canal which contains them.

    " In the same way, when a clot is formed in a vein, the cellular tissue surrounding the vessel is often destroyed by suppuration, before the vascular canal is destroyed itself.   May it not be thus with the œsophagus ?

" A very important point to determine is the period when the abscess appears. It must necessarily vary with the form, volume, and consistence of the foreign body. Yet if we could determine the *minimum* time which elapses between the introduction of the foreign body and the formation of pus, we should then know how much time to allow for expectant treatment.

" We will compare a few cases, although incomplete ones.

I. " An intoxicated man got a chestnut caught in his œsophagus. . He died the nineteenth day. The autopsy showed an abscess communicating with the trachea.

II. " A young man swallowed a turkey bone. He expired the thirteenth day. At the autopsy, the bone was found in an abscess. The œsophagus was perforated.

III. " A lady swallowed the spine of a carp. She died the fourteenth day. Autopsy: a tumor (of pus?) filling the œsophagus. A circular perforation.

IV. " A foreign body in the throat occasioning death. Autopsy: an abscess opening into the chest.

V. " A child of twenty-two months swallowed a bone, and died two months afterwards. Autopsy: the posterior wall of the œsophagus perforated opposite the third cervical vertebra. Caries of the second, third and fourth vertebræ.

VI. " A soldier swallowed a five franc piece. Death at the end of six months. The coin was found

near the cardiac orifice of the stomach. The walls of
the œsophagus were thickened and suppurating.

VII. "A soldier swallowed a fragment of bone,
and died the eighth day, after vomiting blood.
Autopsy: the bone was found opposite the fourth
dorsal vertebra. Its two opposite angles had passed
through the walls of the œsophagus. The lungs on
either side presented a lesion opposite these angles;
and from these points inflammation had radiated to
the other lobes.

VIII. "A man swallowed a beef-bone. Œsophagotomy was performed on the ninth day, and the
bone extracted. The patient died two days afterwards. The autopsy showed the œsophagus perforated in front and behind. A purulent collection extended behind the canal, down to the stomach.

"It results from these observations that, indepen-
dently of lesions of the trachea and great vessels which
environ the œsophagus, foreign bodies in this canal
may lead to the formation of abscess severe enough
to occasion death.

"At what period do these abscesses appear? A
solution of this question would teach us how long we
could wait before performing œsophagotomy without
compromising the success of the operation.

"Now, in M. Demarquay's first case the accident
happened on the 26th of June, and the next day (the
27th) emphysema was observed in the sides of the
neck; perforation of the alimentary canal had already
taken place, and, as it was close to the upper opening of the larynx, the cause of the emphysema is

easy to perceive. The child died the 3d of July, a week after the accident, and we have seen what grave lesions had been produced in so short a time.

" In his second case the accident occurred the 26th of July. Œsophagotomy was performed the 5th of August, the tenth day. At this time the posterior wall of the œsophagus was already perforated, and it is probable that the purulent collection had been formed several days. It is almost certain that the want of success of the operation was owing to this circumstance.

" In the first case which we have taken from other authors, death occurred the nineteenth day, after the formation of an abscess and an opening leading into the trachea.

" In the second, death occurred the thirteenth day, determined by an abscess and a perforation of the œsophagus.

" In the third, death the fourteenth day ; and there was found a tumor in the œsophagus, and a perforation.

" In cases four, five and six, death took place at a later period, but from the same cause.

" In the seventh case, death, which took place the eighth day, was caused by a double perforation of the œsophagus, lesion of the lungs, and pneumonia.

" The eighth case, that of M. Flaubert, of Rouen, bears a striking resemblance to the second one of M. Demarquay. Œsophagotomy was done to extract a bone, which had been in the œsophagus eight days. The patient died two days later, and the œsophagus

was found perforated before and behind, with a purulent collection behind it.

" By these cases we learn that suppuration occurs
very quickly around a foreign body, and that it leads
to death by the end of the first week, or in the course
of the second.

" It is true that there are cases where it has not
occurred until the end of several months, and even a
year, or more ; others, again, where the foreign bodies,
after a considerable time, have come to the surface
spontaneously, and been thus cured ; but such cases
are the exception : and it seems to us to result, from
the cases we have reported, particularly those of MM.
Demarquay and Flaubert, that *the operation, when it
is decided on, should not be deferred beyond the second
or third day.*

" In fact, if death does not occur before one or
two weeks, suppuration precedes it by many days ;
and from the instant it takes place, there is but a
small chance of success.

" This is an important point on which we shall
insist. The death of the patients operated on by
MM. Demarquay and Flaubert, must be referred to a
temporizing policy pursued too long."

In the fatal cases of Arnott and Martini, referred
to in our table, it will be observed that in the first,
the operation was done after five weeks, and pneumonia existed at the time ; in the second, the operation was done on the fourth day, but the pharynx was
already gangrenous.

" Yet, notwithstanding what we have said of the

gravity of perforation and abscess, these complications
are not a positive obstacle to the operation.

"Bégin operated with success the twelfth day,
when there was perforation, and mediastinal abscess.
The wound healed in five weeks.

"Lavacherie operated on the eighth day with
success, when there was perforation, gangrene, and
the foreign body pressing on the common carotid."

Syme, we would add, operated twice, with abscess
—once on the sixth day, and once on the sixteenth.
Both recovered.

" Again, it is not only perforation of the œsophagus
and the formation of retro-pharyngeal and retro-œsophageal abscess which ought to hasten the moment
of operation ; but fatal hæmorrhages, from lesion of
the great vessels, may take place very rapidly.   Thus
in twelve cases reported by Lavacherie, where a
fatal result occurred from arterial lesions, death took
place the sixth, eighth, ninth, and, twice, the tenth
days.

" To determine at what period the surgeon ought
to perform œsophagotomy, M. Demarquay instituted
a series of experiments in order to fix the time when
a foreign body introduced into the œsophagus brings
on perforation, or retro-œsophageal abscess.   To
study the phenomena which occur in animals when
foreign bodies are introduced, we must consider, first,
their size, and second, their angularity ; for the larger
and rougher they are, the sooner they lead to inflammation and perforation.

" Now, in the dogs, into whose œsophagi, by the
10

aid of œsophagotomy, this surgeon introduced fragments of bone, which he fixed there, with a thread
leading outside, the following state of things was observed on the fourth and sixth days.

" The bony prominences were already deeply im-
pressed into the upper parts of the digestive tube,
and little gangrenous points sometimes corresponded
to these impressions. But that which attracted his
attention most was, that *in the cellular tissue around
the œsophagus, a vascular congestion more or less vivid
was seen, and, sometimes, pus, without the œsophagus
being perforated.* The continued presence of the
foreign body was sufficient to bring on these disorders,
which we can easily understand when we reflect on
the vascularity of the digestive canal at this point,
the looseness of the surrounding cellular tissue, and,
above all, the epithelial layer which coats the inner
surface of the œsophagus. Another noteworthy
circumstance is that in animals, as in men, rough,
foreign bodies arrested in the œsophagus bring on
acute pain, and check the taking of food, both
circumstances fovorable to the formation of pus.

" For all these reasons we would agree with the
Surgeon of *Maison Municipale de Santé,* in advising
an early opening of the œsophagus, when the foreign
body is large, or brings on severe pain, or renders
the taking of food difficult or impossible. If we delay
beyond the fifth or sixth day, we run the risk of finding a retro-œsophageal abscess ; and we have seen all
the gravity of this complication.

" As to the difficulties of the operation it is not to

be denied that they are great, and that none but an experienced hand ought to undertake them. We need not add that this truly delicate operation should not be tried until we have made, with prudence, all the attempts at extraction recommended in such cases."

These attempts at extraction, we may be allowed to add, are capable of infinite mischief. In order to pass even a stomach tube into the œsophagus, we are obliged to keep close to the posterior wall. This wall has been perforated, even by a probang. Owing to its structure and connections it is especially prone to such accidents. United by loose connective tissue only to the prevertebral muscles, those who have made the *coup du pharynx* in the amphitheatre know how easily it can be separated. On this account it is readily pushed before the instrument, and perforated, or at least injured to the degree of bringing on ulceration, unless the greatest care be used. It is in this respect analogous to the membranous portion of the urethra, and should be handled as gently. We think that it is not commonly so regarded; and that greater force is often used here, than would be thought justifiable in the other mucous canal.

Among the least objectionable of the instruments devised to withdraw foreign bodies are the soft sponge probang, and the snare, composed of many loops of thread.*

* We desire to call attention to two improved forms of œsophageal probang, one by Dr. David Rice (Boston Medical and Surgical Journal, No. 8, Vol. 76); the other, an East Indian invention, revived by Dr. Sayre (New York Medical

The former, when used to push down foreign
bodies, is introduced moist and expanded ; but when
it is designed to withdraw the object, it is passed
down dry and compressed, made to swell by drinking
some water, and then drawn up, pressing the foreign
body before it.

A similar measure might be employed by passing
down a collapsed rubber bag, with a tube attached,
and inflating it before withdrawing.

We must reflect that we do not know how the
foreign body lies ; and that if it be sharp and angular,
it is much more likely to be imbedded by any of these
means, than to come up, or go down.   The same
objection applies to emetics.

In comparing the chances of nature or of art in
removing a foreign body from the pharynx or œsophagus, it is fair to say that although many such in-
truders are gotten rid of safely by the former, yet
very many also lead to a fatal result.

Constitutional irritation and inanition are prompt
and firm supporters of suppuration; and they are
always ready to act when a foreign body is retained
in the throat beyond a few hours.   Such retention,
as we have seen, may lead to pressure on the air-
passages and asphyxia ; to abscess leading to the

Journal, No. 3, Vol. 7). The first consists of a sponge probang, with *conical
sponge*, base uppermost, which unfolds and sweeps the œsophagus on being
withdrawn. The second is made of *bristles*, which passing down straight, are
expanded like an umbrella on being withdrawn. Both are extremely ingenious
and useful. With their aid, and that of the *œsophagoscope*, we may hope to reduce
even more the already few and rare cases in which the operation of œsophagotomy becomes necessary.

same accident, mechanically; to spasm of the glottis; to ulceration and perforation of the œsophagus, into the trachea, the blood-vessels, the bronchi, the mediastinum; while the pus of a retro-pharyngeal abscess may burrow rapidly down, unobstructed, into the pleura, or end in caries of the intervertebral substance. Suppuration also may take place in the cellular tissue around the œsophagus before perforation be accomplished, as shown by the experiments on the lower animals, and as indicated by the rapid suppuration in our first case. Such suppuration may become diffuse, and end in gangrene.

In view of all these perils why should not œsophagotomy be the rule, after reasonable attempts at extraction have failed, just as an operation is the rule in strangulated hernia, after reasonable attempts at taxis have failed?

We only lose by delay. The experiments of Demarquay have proved that suppuration is imminent if we wait longer than the third or fourth day. In the whole seventeen cases we find only four deaths, or less than twenty-five per cent. And in every one of these, death was due to secondary complications; due either to delay, or to overtreatment in attempts at extraction. In one, there was pneumonia; in two, gangrene; in the remaining one, abscess. No projection externally of the foreign body need be waited for, or expected. As to the manner of the operation, we have given our reasons for the lateral method, which, indeed, is favored by most writers. We need not remind the anatomist, that the nerves are very constant in their distribution, and can all be avoided. And if anomalies of arteries are feared, there is but one of much consequence, and that very rare, namely, the origin of the right subclavian from the arch of the aorta, in which case it crosses behind the œsophagus.

In comparison with the perils of expectant treatment in surgery, we are almost ready to say, that no dangers from the knife, in an educated hand, can equal those of delay.

THE following table comprises all the cases of œsophago-tomy we have been able to find. Others may exist. But it will be noticed that we have collected more than double the number given in the article on this operation by Mr. Henry Gray, in Holmes's System of Surgery.

We are indebted to our friend, Dr. L. Voss, of New York, for a complete revision of our table; for a comparison of authorities, and for several additional cases.

References to authorities consulted, and to cases quoted, will be found on the pages following the Table.

TABLE OF CASES OF ŒSOPHAGOTOMY.

POINT OF IMPACTION.	TREATMENT BEFORE OPERATION.	OPERATION, WHEN PERFORMED.	RESULT.
Œsophagus; where, not stated; could be felt outside.	Attempts to push it down	Not stated.	Recovered.
Not stated.	Not stated.	Not stated.	Recovered.
Œsophagus; lower part of neck.	Touched the foreign body; attempts to dislodge it.	Operation twelfth day, left side.	Speedy recovery.
Œsophagus; lower part of neck.	Touched the body; every means tried to dislodge it.	Operation eighth day, left side.	Recovered.
Lower part of pharynx.	Emetics, and various attempts to dislodge it.	Operation after five weeks, on right side.	Death fifty-six hours after operation.
Œsophagus—perforation of; lying on carotid.	Not stated.	Operation eighth day.	Recovered.
Could be felt outside, projecting above clavicle.	Bleeding, tartar emetic in veins, belladonna enemata, and sixty attempts with instruments.	Operation fourth day; bono swallowed.	Death two days after operation.
Pharynx; tail seen in fauces.	Vain attempts to withdraw through mouth.	Operation after several days.	Recovery in six weeks.

				Junction of pharynx and œsophagus. No external projection.	Attempts at withdrawal with forceps; emetics.	Operation fourth day, left side.	Recov. in 4 w.; permanent alteration of voice	
12	1856	M	Gold tooth-plate containing a false incisor.	Junction of pharynx and œsophagus. No external projection.	Attempts at withdrawal with forceps; emetics.	Operation fourth day, left side.	Recov. in 4 w.; permanent alteration of voice	Cock.
13	1861		Thin piece of mutton bone one inch sq.	Œsophagus; no external projection.	Could not be touched by fauces.	Operation sixth day.	Recovery in two weeks.	Syme.
14	1862		A coin.	Opposite top of sternum.	Coin touched by bougie.	Operation after two months.	Recovery; swallowed in a week.	Syme.
15	1863	M	Bone.	Not stated.	Not stated.	Not stated.	Recovered.	Fourier.
16	1864	F	Peach stone.	Not stated.	Not stated.	Not stated.	Recovered.	Arnold.
17	1866	M	Cod-fish bone.	Junction of pharynx and œsophagus. No projection.	Vomiting; exploration by finger and probang; rigors.	Operation third day, right side.	Recovered.	Cheever.
18	1866	M	Brass pin.	Below top of sternum. No projection.	Vomiting; long probang.	Operation third day, left side.	Recovery in five weeks.	Cheever.
19	1867	M	Tooth-plate.	Opposite left cricold.	Various explorations.	Operation third day.	Recovered.	Cock.
20	1867	F	Brass pin.	Apparently opposite left cricold.	Attempts during four months.	Operation after four months.	Recovered.	Hitchcock.
21	1868	F	Supposed to be a pin.	Junction of pharynx and œsophagus.	Various attempts.	Operation after eight months.	Recovered.	Cheever.

Foreign bodies:—Authentic cases, 21 ; Deaths, 4 ; Recoveries, 17.

NOTE. In cases 20 and 21 no foreign body was found. The lapse of time (four and eight months after the swallowing of the pins) may have favored their escape, or becoming encysted outside the œsophagus. For the severity of the symptoms the reader is referred to the history of the cases.

FOR STRICTURE.

No. 22.	1844.	Male.	Stricture.	Died.	Watson.
23.	1845.	Male.	Stricture.	Died.	De Lavacherie.
24.	1864.	Male.	Stricture.	Died.	Bruns.

Operations for stricture, 3.   Deaths, 3.

No. 25. Obscure allusion to the case of a beer-cork in the throat, in Portea ; neither operator, details nor result given. [Vide Eve's Surgical Cases.]

[See references on next page.]

REFERENCES.

Cases.

No. 1. Mémoires de l'Acad. de Chirurgie, Tom. III., p. 10 (Edition 1819, 8vo.).

" 2. *Ibid.*

" 3. Journal Universel et Hebdomadaire de Med. et de Chirurg., 1833, Avril 20, p. 94.

" 4. *Ibid.*

" 5. Med.-Chirurg. Transact., Vol. XVIII., p. 86.

" 6. Mem. de l'Acad. Royale de Med. de Belgique:Bruxelles, 1845. Journal de Chirurgie, par Malgaigne, 1845, Tom. III., p. 337.

" 7. Wurtemberg Correspondenz Blatt, 1844. Journal de Chirurgie, par Malgaigne, 1845, Tom. III., p. 336.

" 8. London Lancet, 1854, Vol. II., p. 260. (From Ceylon Miscellany, Vol. I., No. 2.)

" 9. Gazette des Hôpitaux, 1857, No. 88.

" 10. Gazette des Hôpitaux, 1854, p. 400. Gazette Hebdomadaire, Nov., 1861.

" 11. Syme's Clin. Surgery, 1861, p. 94.

" 12. Guy's Hospital Reports, 3d Series, 1858, Vol. IV., p. 217.

" 13. Syme's Clin. Surgery, 1861, p. 91.

" 14. Edinburgh Journal, 1862, p. 1010.

" 15. Gazette des Hôpitaux, Févr., 1864.

" 16. F. C. Morgagni, VI., 1864, p. 352.

" 17, 18. Cheever's Monograph, 1868.

" 19. Guy's Hospital Reports, 3d Series, Vol. XIII.

" 20. Boston Medical and Surgical Journal, July 16th, 1868.

" 21. Cheever's Monograph, 1868.

" 22. Dublin Journal, May, 1845.

" 23. Bulletin de l'Acad. de Med. de Belgique, 1845. Journal de Chirurgie, par Malgaigne, Tom. III., 1845, p. 371.

" 24. Deutsche Klinik, 1865, p. 37.

" 25. Eve's Surgical Cases.

REFERENCES TO WRITERS OF MONOGRAPHS; OR SPECIAL ARTICLES.

Berlinghieri. Florence, 1780.

Hévin. Mémoires de l'Academie de Chirurgie, Tome I., 1819.

Verduc. Pathologie Chirurgicale.

Guattani. Mémoires de l'Academie Royale de Chirurgie, Tome III.

Bégin. Journal Hebdomadaire, 1833.

Hébra. British and Foreign Medical Review, Vol. XVI., page 53.

Lavacherie. Mémoires de l'Academie Royale de Belgique, 1844.

Créquy. Gazette Hebdomadaire de Médecine et Chirurgie, Nov., 1861.

Arnott. Medico-Chirurgical Transactions, Vol. XVIII.

Cock. Guy's Hospital Reports, Vol. IV., 1858.

Gray. Holmes's Surgery, Vol. II., 1861.

Adelmann. Prager Vierteljahrschrift für die practische Heilkunde, Vol. 96, 1867, p. 107.

AUTHORS CONSULTED, OR QUOTED.

Boyer, Cooper (Surg. Dict.), Charles Bell, Hennen, Velpeau, Fergusson, Syme, Nélaton, Sédillot, Guérin, Erichsen, Holmes, Gross.

VELPEAU'S LESSONS

UPON THE

DIAGNOSIS AND TREATMENT

OF

SURGICAL DISEASES.

COLLECTED AND EDITED BY

A. REYNARD.

Translation by W. C. B. FIFIELD, M.D.

1 Vol. English cloth, $1.00. JAMES CAMPBELL, 18 Tremont Street, Boston.

"THIS modest little book contains a statistical *résumé*, by the author, of his surgical experience in the hospital wards under his care during the year. He treats his subject under the successive headings :—Generalities, Fractures, Affections of the Joints, Inflammation and Abscesses, Affections of the Lymphatic System, Burns and Contusions, Affections of the Genito-Urinary Organs, Affections of the Aural Region, Affections of the Eyes, Statistics of Operations. We have a special liking for such works, which give us the most authoritative opinions of the elders of the medical profession, who have reached the time when the judgment is least biased by the rivalries and personal influences which are so apt to mislead younger minds. It is of vastly more value than many more ambitious and bulky Works."—*Boston Medical and Surgical Journal.*

"He not unfrequently surprises us by the simplicity of his expedients for the aid of ' Nature in Disease,' and rarely, if ever, fails in making out his case ; as a whole, the work is not only instructive but entertaining, and may be regarded as one of our landmarks of minor surgery, upon our skill in which much of our success will be found to depend."—*Medical Record.*

"It is rare that so small a book contains so many suggestions of great practical worth, and throws so much light on certain debated points, as Velpeau's Lessons. Though nominally a review of one year's practice, it is in reality an epitome of the experience of a lifetime."—*Detroit Review.*

"All who value the teachings of this great man will not lose the opportunity of obtaining them, as presented in this brief and economical form."—*Richmond Medical Journal.*

METHOMANIA:

A TREATISE ON ALCOHOLIC POISONING.

BY

ALBERT DAY, M.D.

SUPERINTENDENT OF THE NEW YORK INEBRIATE ASYLUM, BINGHAMTON, N. Y.

WITH AN APPENDIX BY

HORATIO R. STORER, M.D.

1 Vol. 16mo. English cloth, $0.75. Published by JAMES CAMPBELL, 18 Tremont Street, Boston.

THIS little volume is a treatise on those poisons which first stimulate, then narcotize. The author has had ample means to thoroughly investigate his subject, having had over two thousand cases under his care, a great many of which were suffering under one of the different forms of *mania à potu.*

Dr. Storer says : " I take pleasure in acknowledging alike the excellence of the treatise he has written, its strictly philosophical spirit, the practical influence it must have upon the community, and, above all, the truly scientific manner in which ebriety and its effects, almost for the first time in the history of medicine, are now being treated at his hands."

www.ingramcontent.com/pod-product-compliance
Lightning Source LLC
Chambersburg PA
CBHW021955190326
41519CB00009B/1275